I0105742

ANCIENT GREECE: SHIPPING AND TRADING LESSONS FROM HISTORY

By; Mustafa Nejem

CONTENTS

INTRODUCTION

Undoubtedly, ancient Greek mythology, with its powerful figures like Zeus and the captivating tale of Hades and Persephone, is renowned worldwide. The splendour of their architecture, the artistic finesse of pottery and everyday objects, and the enduring literary masterpieces like the Iliad and the Odyssey, coupled with the profound philosophies of Aristotle, Plato, and Socrates, depict ancient Greece as much more than just a civilisation. It stands as the epitome of learning and excellence in nearly every facet of life, leaving a deep mark on the course of trading history.

However, there are lesser-known but equally valuable lessons to be learned from the marvels of ancient Greece. Their exceptional trading and shipbuilding skills were instrumental in shaping their civilisation. With a wealth of coastline and numerous city-states, ancient Greece fostered a vibrant maritime trading system. This book will delve into their lives as remarkable shipbuilders and traders who ventured offshore, propelling their economy with ingenious strategies. We will also explore the cultural impact of these aspects and the enduring lessons modern society can draw from their achievements.

Ancient Greece were excellent and witty traders, recognising the importance of a robust navy not only for military purposes but also for political and trading dominance. They implemented a range of strategies to protect their trade routes, including convoy escorts, blockade patrols, coastal fortifications, preemptive raids, and control of critical maritime passages. Their advanced warships, such as the trireme, served as highly effective deterrents against piracy and bolstered their position as dominant maritime powers.

Another captivating facet of ancient Greek trade was the utilisation of amphoras. These containers were employed to transport various goods, including wine, oil, and grain, and played a pivotal role in categorising and standardising products. It's remarkable to envision the existence of standardised packaging during the time of ancient Greece. However, these amphoras weren't just practical containers; they were also exquisite works of art that were sought after as products in their own right.

Amphoras played a crucial role not only in international trade but also in local exchanges. However, for local trade and as bustling hubs where international traders also convened, the agoras took centre stage. Agoras were multifunctional spaces for selling goods, hosting meetings, engaging in political discussions, and even conducting religious rituals. These agoras held immense political and economic significance in ancient Greece, and we can confidently draw parallels between these ancient hubs and the modern concepts of malls and markets. Ancient Greece's agoras laid the foundation for our contemporary commercial marketplace.

Greek traders engaged in local and international trade and established an extensive network of trade routes and colonies. These colonies were strategically situated in regions like the Mediterranean, the Black Sea, and neighbouring areas to address overpopulation, secure valuable resources such as timber, metals, and agricultural products, and mitigate the risk of civil conflicts.

Greek merchants played a pivotal role in international trade by acting as middlemen, importing essential goods and exporting Greek products through these trade routes and colonies. Moreover, these colonies served as conduits for spreading Greek culture, language, and religion to other regions.

The rich and diverse Greek culture, marked by the fusion of artistic beauty and natural elements, was evident in their architecture, sculptures, and mosaics. This cultural richness not only enhanced trade but also facilitated cultural exchanges that contributed to the broader influence of ancient Greek civilisation.

Indeed, ancient Greece held the honour of introducing the first-ever coinage system in history. Greek coins had a profound impact on trade by offering a trusted and standardised medium of exchange. Foreign civilisations highly coveted Greek coins, particularly those minted in Athens, because of their exceptional quality and reliability. These coins were pivotal in facilitating trade and economic transactions across ancient Greece and beyond.

At last, ancient Greek shipbuilders made revolutionary strides in naval engineering, particularly with the development of the powerful trireme warship, which can be traced back to the legacy of the Phoenicians, the great shipbuilders of their era. These innovations not only transformed the landscape of naval warfare but also left a profound imprint on the broader evolution of seafaring vessels and the world of trade. Ancient Greece's contributions to ship design and technology had a lasting impact on maritime history.

Overall, the legacy of ancient Greek traders and shipbuilders has left an indelible mark in history. Their achievements continue to offer valuable lessons and inspiration for modern businesses and maritime engineers, encouraging innovation while preserving the rich traditions inherited from the past. The lessons learned from the history of ancient Greece serve as a timeless source of wisdom for shaping and advancing modern practices and designs.

Chapter 1

NAVAL POWER IN ANCIENT GREECE

Imagine the historical global, where the Mediterranean Sea becomes like a limitless blue motorway. The Greeks lived in a world surrounded by the sea. They quickly found out that having a robust military became crucial. Their adventure through time teaches us so much. It's now not just about combat; it is about defending their freedom, making their subculture regarded by others, and becoming a powerful pressure inside the historical international. One of the important things the Greeks did was to increase techniques to defend their ships and exchange routes. They had warships that acted like bodyguards for service provider ships, ensuring pirates did not attack them. They also knew how to manipulate crucial spots within the sea.

Therefore, this chapter will take a deep insight into ancient Greece's naval strength. We'll be looking at the role of naval power in Greek Trade, its historical significance and the lessons that modern businesses can learn from Greek Naval Power. We'll also delve into the strategies for protecting the trade routes.

Historical Context of Greek Naval Strength

The precursors of Greek naval dominance lie in the maritime kingdoms of the Minoans and Mycenaeans during the Bronze Age from 3000 BC to 1100 BC. The Minoans were daring seafarers based on the island of Crete who conducted flourishing oceanic trade using fast galleys manned by skilled oarsmen. Their merchant ships crisscrossed the Mediterranean, carrying Minoan wares from Mesopotamia to Iberia. They transported elaborately painted pottery, bronze weaponry, golden jewellery, and rich textiles across their trade networks. Minoan ships hugged coastlines to hop along trading posts and ventured into open waters to reach major coastal cities.

Their naval reach stretched as far south as Egypt, where they exchanged precious metals for exotic painted glass, ivory carvings, and papyrus scrolls. The Minoans maintained a sphere of commercial dominance through naval power, suppressing piracy, and protecting seaborne trade routes. Their ships were engineered for transportation and tactical manoeuvrability, with tapered hulls and angular rigging capable of tacking against headwinds. The later Mycenaeans absorbed and expanded upon Minoan nautical expertise. From the mainland Greek kingdoms, they extended trade and naval control across the Aegean to islands like Crete and Rhodes.

Mycenaean seafarers established colonies and trading posts as distant as Italy, Sicily, Cyprus, and the Black Sea. Raw materials like copper, tin, and gold were shipped back to

Greece to support the ambitions of Mycenaean rulers, who constructed imposing palace complexes financed through maritime trade, including towering citadels at Mycenae, Tiryns, and Pylos. Minoans and Mycenaeans were pioneering naval engineers, incorporating shipbuilding innovations like tapered hulls, angular rigging, and tiered rowing configurations learned from Egyptian and Levantine maritime powers to maintain regional naval dominance during a turbulent era.

This rich legacy of seaborne commerce provided the bedrock for the city-states emerging on Greek coasts and islands from the 8th century BC onwards. Naval powers like Corinth, Aegina, and Rhodes quickly built formidable fleets that unlocked the potential for widespread colonisation and trade across the Mediterranean. But Athens leveraged naval might to become the preeminent Greek sea power by the 5th century BC. Through maritime commerce and naval conquest, Athens brought the islands of the Aegean and the eastern Mediterranean under political and economic control. Athenian triremes were the most advanced warships of their era, manned by skilled sailors and rowers operating in coordinated formations. This naval supremacy allowed Athens to defend itself against multiple Persian attempts at invasion and expansion. At the decisive Battle of Salamis in 480 BC, Athens lured the gigantic Persian fleet into the narrow Salamis straits and won a crushing victory, cementing Greek autonomy in the Aegean. For decades after, Athens maintained naval dominance over rivals like Sparta, forging a prosperous Athenian Empire of over 150 tributary city-states around the Aegean. Even in defeat during the Peloponnesian War from 431-404 BC, Athenian naval power proved resilient, demonstrating its vital role in the fate of the Greek city-state system. This legacy of maritime strength would shape Mediterranean geopolitics and the course of Western civilisation for centuries.

The Role of Naval Power in Greek Trade

The extensive maritime trade networks that connected the Greek world relied heavily on the protection and control afforded by dominant naval fleets. Athens' formidable navy allowed it to dominate commerce throughout the Aegean and Black Sea regions for decades, safeguarding vital trade routes from pirates while weakening economic rivals through naval blockades of ports.

This maritime commerce generated immense wealth for Athens, funding further expansion of its navy while transforming the city into a cosmopolitan centre for industry, finance, and culture. Athens' port of Piraeus became the central hub for trade in the entire Mediterranean under Athenian dominance from the 5th to 4th centuries BC. Athenian silver owl coins were widely used as currency. An empire of Greek trading colonies spread across the Mediterranean shores and Black Sea outposts.

Athens demanded tribute from other Aegean city-states in return for the naval protection it provided. But this naval empire bred resentment, with allies rebelling when given the chance, as during the Social War in 357 BC against Athenian overreach. Yet Athens' economic and cultural pre-eminence highlighted the substantial benefits of leveraging naval supremacy to dominate maritime trade. Its eventual defeat in the Peloponnesian War demonstrated the perils of overextension.

Naval Warfare's Decisive Impact

Naval power was a determining factor in conflicts between Greek city-states. Sea battles decided the fates of empires and alliances. The Athenian victory at the Battle of Salamis in 480 BC preserved Greek independence against Persian encroachment. Repeated Persian attempts to subjugate Greece ultimately failed, largely due to Greek naval strength. Naval clashes between Athens and Sparta influenced the course of the Peloponnesian War, with a crushing Athenian defeat at Aegospotami in 405 BC, sealing the end of an era of Athenian dominance. The ability to project naval force, disrupt enemy seaborne supplies, and maintain lines of communication and logistics through maritime channels often proved decisive in Greek interstate warfare on both land and sea.

Naval power allowed Athens to suppress revolts across its empire, from Samos to Lesbos. When its navy faltered, rebellions flared across Athenian holdings. The fates of cities like Syracuse, Corcyra, and Byzantium hinged on controlling their ports and sea approaches. Naval warfare was risky since losing a fleet could weaken a city's economy and defences in one stroke. But the rewards for the victors were equally substantial, as Athens demonstrated for decades.

Lessons for Modern Businesses from Greek Naval Power

The historic Greeks, often considered masters of the ocean, offer a treasure trove of instructions for current corporations. Their maritime legacy gives valuable insights into the significance of naval energy and gives enduring ideas that resonate across the centuries. Let's delve into those lessons and explore their present-day relevance.

The Power of Maritime Dominance

The ancient Greeks diagnosed that manipulation over maritime routes turned into instrumental for shielding political independence. This lesson extends to the present-day enterprise global, where controlling crucial supply chains and distribution networks may be a source of aggressive gain. As town-states defend their sovereignty through naval electricity, businesses must steadily defend their strategic belongings and sources.

Economic Prosperity and Trade

Naval supremacy facilitated the establishment of flourishing change networks within the historical world. In the trendy global financial system, groups have to admire the role of efficient logistics, transportation, and delivery chains in fostering economic increase. The capability to connect to global markets and steady exchange routes can drive financial fulfilment.

The Cultural Ripple Effect

The Greeks simply failed to dominate through naval strength; they disseminated their values, institutions, and traditions across the Mediterranean. This lesson underscores the significance of brand picture and popularity in the enterprise world. Successful businesses provide products or services and export their identity and values, enriching the cultures they touch.

Enriching Mother Cities through Trade

Greek colonies enriched their mom towns through trade, developing collectively useful relationships. Cutting-edge corporations, subsidiaries, franchises, and international enlargement can contribute to the growth of the discern business enterprise. The network effect, which is controlled effectively, can bring about a more potent and assorted business environment.

Continuous Investment in Superior Resources

The Greeks understood that keeping naval supremacy required continuous investment in advanced warships and a professional crew. Similarly, contemporary organisations must prioritise ongoing investment in studies and development, era, and human capital. Corporations must be dedicated to innovation and the pursuit of excellence to live aggressively.

Preparedness for Contingencies

Just because the Greeks remained organised for capability threats even during peace, agencies must preserve their readiness for contingencies. This consists of strategic planning, change management, and disaster preparedness. Companies must be agile and attentive to the rapidly changing marketplace and unexpected demanding situations.

Strategic Alliances and Collaborations

The Greek metropolis-states often shaped alliances for mutual defence. In the corporate international, strategic alliances and collaborations can become powerful tools. Businesses can leverage partnerships, joint ventures, and collaborations to access new markets and proportionate assets and gain from every other's understanding.

Respect for Tradition and Innovation

The ancient Greeks constructed upon centuries of accumulated knowledge at the same time as embracing innovation. This balance between culture and innovation is relevant to trendy agencies. Companies should appreciate installed practices and middle values whilst adapting to new technologies, markets, and consumer expectations.

Sustainability and Environmental Responsibility

The maritime instructions from the Greeks took place at a time when environmental attention was no longer a global concern. Today, corporations ought to include sustainability into their core practices. Environmental duty, smooth technologies, and sustainable operations are vital for lengthy-term success and shielding our planet.

The Role of Ethical Leadership

The Greeks valued leadership and selection-making in their maritime endeavours. Modern agencies need moral management to guide their moves, make responsible picks, and ensure long-term viability. Ethical leadership builds belief, fosters corporate duty, and creates recognition.

Global Connectivity and Cooperation

Global exchange and connectivity were crucial elements of the Greek maritime legacy. Modern agencies must include worldwide interconnectedness, cooperate with companions worldwide, and contribute to global collaboration. Operating in a global economy requires expertise in various cultures, markets, and regulatory environments.

Strategies for Protecting Trade Routes

The ancient Greeks, masters of naval power, used various naval tactics and technology to ensure the steady flow of their far-reaching trade networks. In a world where piracy and war threatened merchant ships, these methods protected economic prosperity and national security.

Naval Convoy Escorts

Warships often went along with merchant ship convoys on major trade routes. This provided safety in numbers, discouraging pirates and protecting valuable cargo. In recognising this tactic's importance, Athens maintained dedicated squadrons for convoy protection.

Blockade Patrols

Allied city-states often worked together by contributing warships to form blockade patrols in high-risk areas, like the treacherous Straits of Corinth. This team effort successfully suppressed piracy, sending a clear message to potential raiders that the Greek city-states were united against any threats to their maritime interests.

Coastal Fortresses

Heavily fortified watchtowers and walled coastal forts dotted the landscape, guarding the bays and islands used as safe harbours by merchant fleets. When warships were unavailable, these fortifications provided critical protection, serving as deterrents to would-be attackers and safe havens for trade vessels.

Pre-emptive Raids

Fast naval squadrons undertook preemptive strikes on known pirate dens along coasts and islands to disrupt piracy at its source. These missions aimed to destroy pirate ships and infrastructure, effectively ending their operations in the area. The Greek navy's proactive approach struck fear into the hearts of pirates and reduced their presence.

Maritime Chokepoints

Strategically positioned naval patrols bottlenecked passages vital to maritime trade, including the Straits of Corinth. By controlling these chokepoints, the Greeks blocked pirates from swarming into critical sea lanes, effectively safeguarding their trade routes.

Advanced Warships

The iconic Greek trireme played a pivotal role in securing the safety of trade convoys and blockade patrols. These warships were known for their agility, speed, and manœuvrability. Manned by skilled oarsmen and marines, triremes were formidable deterrents to most pirates. Their presence alone often dissuaded would-be attackers from pursuing a life of piracy.

Naval Blockades

During conflicts between Greek city-states, naval blockades of enemy ports with squadrons of triremes were commonplace. This tactic aimed to weaken the adversary's trade and deny them access to critical supplies. Athens, in particular, leveraged this tactic effectively to gain the upper hand during wars.

Cultural Integration

The Greek approach went beyond mere naval might. By establishing colonies and trade centres across the Mediterranean, they created a far-reaching network of Greek culture. The protection of local naval patrols underpinned this cultural integration. Greek influence spread not only through force but also through the dissemination of their way of life and values.

 In conclusion, the ancient Greeks demonstrated over centuries how naval dominance could enable the growth of commerce, culture, and civilisation across a broad maritime region. Their mastery of shipbuilding, maritime trade, and naval warfare left an enduring legacy of seapower's pivotal role in the rise of empires and economies. Continuous investment in technological advances enhanced Greek naval might, while widespread cultural networks strengthened political bonds and promoted shared security. Checking piracy and enemy action through interlocking naval patrols, coastal fortifications, and preemptive raids suppressed threats to prosperity.

Their achievements and failures provided enduring models for great maritime powers throughout history seeking to understand the close linkage between seapower and civilisation. Naval dominance enabled the Greeks to spread their culture but also sowed the seeds of ruinous conflict when unchecked by moderation. This complex legacy would echo through the rise and fall of maritime empires from Rome to Britain, framing the bittersweet fruits of nautical supremacy balanced against the gravity of hubris.

Chapter 2

ANCIENT GREECE AMPHORAS AND STANDARDIZATION

In Ancient Greece, there were various trade networks in which Amphoras were used to exchange different commodities. They were of different types and were important for categorizing the goods in the ancient Greek world. The Ancient Greeks have always used beneficial measures to transport goods to distant lands. The Amphoras were used along with the clear packaging standards that ensure uniformity. All the packaging and containerisation were very suitable for promoting sustainability. So, modern businesses should also implement different packaging practices and standardisation techniques to get efficiency and transparency.

Importance of Amphoras in Trade and Commerce

Amphoras have played a very important and significant role in the trade and commerce. In Ancient Greece they were essential for the society and the economy of ancient Greece. They were ceramic containers that were designed for transporting and storing numerous goods. These goods include vinegar, oil, and wine. They were a great source of secured transfer of valuable commodities over long distances. Amphoras were produced in different standardised shapes and sizes. The standardising of this amphora and containers promoted reliability and trust in trade and commerce. They also facilitated trade within Greece and through the Mediterranean region.

The Greek merchants were involved in goods exchange with the other ancient civilisations. It helped the people of Greece in the cultural exchange and growth of trade networks. For the Ancient Greek economy, trade was a vital component, and Amphoras played a central role in boosting economic activities. The transportation, production, and exchange of goods using the different types of containers and Amphoras created many economic opportunities for the people of Greece. These vessels were also used for artistic expression and showcasing the craftsmanship of the Greek potters.

Examples of Goods Transported in Amphoras

Greek wine was a very valuable commodity, and amphoras were highly preferred vessels for storing and transporting it. The shape and packaging of the amphora helped preserve the quality of the wine during the transport. Olive oil was also another valuable commodity that was transported through amphoras. These containers protected the olive oil from air and light exposure. This helped in preventing it from spoilage.

Grains were also traded through amphoras, including barley and wheat. The design of Amphora used for transporting grains ensures that they remain protected from pests during transit. The people of Greece in ancient times also used these vessels for transporting fish sauce. It was a very common ingredient in different cuisines at that time. So, the fish sauce was also traded and transported through the amphoras.

Types of Amphoras in Ancient Greece for Trade and Commerce

In ancient Greece, numerous Amphoras were used and designed for specific purposes. In commerce, these amphoras were distinguished by their size, design, shape, and the type of goods they contained.

Dressel 1 was used for storing and transporting wine and olive oil. These were large and elongated with two handles and a narrow neck. They were available throughout the Mediterranean region. These vessels were very helpful in the transportation of oils very safely without any leakage.

Dressel 2,3,4 containers were used for transporting different types of liquid. It includes wine and olive oil. They had more bulbous bodies as compared to Dressel 1 amphora. They were of medium size and were used for safe transportation.

Moreover, there was another amphora type known as the Knidian Amphora. It comprised a long neck and two small handles. These containers were used for transporting and storing wine. But these vessels were only associated with Knidian's region in Asia Minor. So, they were mainly used in the trade of wine in the particular region.

Another type of amphora that was used in ancient Greece was Rhodian amphora. They were tall and slender with a long neck. They also comprised two handles and were used for transporting wine to the island of Rhodes.

These different types of amphoras were very helpful for categorising and transporting goods in the ancient Greek world. Their design, style, shapes, origins, and markings provide very valuable information to the archaeologists. They were very helpful for the historians in the ancient trade networks for exchanging commodities between regions.

Benefits of Efficient Handling Storage and Transport

The efficient handling, storage, and transport of goods in ancient Greece offers numerous benefits and lessons for modern businesses. The streamlined processes are useful for reducing the operational cost. It is also helpful in minimising the risk of product leakage, product damage, or any other factor leading to the loss of efficient logistics. It will also result in quick order fulfilment and timely delivery. This leads to improved customer satisfaction and retention. Moreover, optimised storage practices are also used, which will minimise the excess inventory and reduce the cost of storage and transport.

Proper handling and storage will also prevent the contamination of goods, ensuring that goods are maintained in high quality. So modern businesses will easily excel through efficient transport and storage by gaining a competitive edge in the market. Similarly, efficient logistics will also help contribute to sustainability. So, the importance of efficient handling, storage, and transport enables different modern companies to meet customer expectations and market demands quickly.

Challenges in Implementing Standardisation in Amphoras

Implementing standardisation in containerisation and packaging will be a valuable practice. But it will include several challenges. The stakeholders and employees can resist the changes for specialised packaging and containerisation. So, overcoming resistance and ensuring the purchase from all the stakeholders will be challenging. Moreover, different industries may have unique packaging needs. So, achieving broad standardisation that meets the industry's specific requirements will be complex. Also, the standing containers and vessels may not require significant investments in processes, materials, and equipment.

So, these costs of transitions will be very challenging steps in the transfer and storage of goods. Additionally, the existing infrastructure and the supply chain management may differ from the standardised containers. So, it will also destroy the supply chain management. The regulatory requirements, including safety and labelling standards during the standardisation implementation, will be challenging, as well as standardised containers unsuitable for all types of goods. So, the heavy and oversized items will require special handling, which can cause complications in the standardisation process. Also, ensuring that all the stakeholders and partners are informed about the standardised process is a big task.

Despite these challenges, the benefits of standardisation of the containers are more successful. Implementation requires careful collaboration, planning, and understanding of customer needs and demands.

Examples of Industries in Ancient Greece that Have Faced Challenges

In Ancient Greece, numerous Industries faced challenges and the implementation of standardization. The challenges were due to the limitations of resources, technology, and communication in the early times.

Pottery and Ceramics:

One of the most notable examples is the ancient Greek pottery and ceramics. They were widely used in terms of decoration and were available in a variety of sizes. This variety of pottery and ceramics made standardisation challenging. This was only due to the limited access to clay sources and diverse art preferences. It was also due to the absence of modern production techniques that was the main hindrance to standardisation. In ancient Greece, using pottery and ceramic items, mainly amphoras, was highly challenging to standardise.

Currency and Coinage:

Another example includes the use of currency and coinage. Ancient Greece had various states and cities that had their coins and currency with different designs and weights. Here, the problem arising in the standardisation of currency was local autonomy and the lack of centralised authority. So, trade was an important requirement for dealing with different coin systems and currencies.

Textile Manufacturing:

Ancient Greek systems also comprised different textiles, including fabrics and clothing, that differed in style, design, materials, and weaving techniques. There was unlimited access to the required materials to produce cloth and fabrics. Also, craft production was available. These two factors were the main ones preventing standardisation in ancient Greece's textile manufacturing.

Ship Building:

Additional Greek shipbuilders have also constructed different types of ships. They were used for different purposes according to their sizes and design. The lack of standardised materials for the construction of ships and the transportation of goods contributed to the lack of standardisation of the vessels.

Construction:

The construction methods have been varied in different states and cities of Ancient Greece. It resulted in different building techniques and styles in the past. But at that time, the construction materials, local transitions and architectural preferences caused the absence of standardised construction practices. These challenges were gradually overcome in Ancient Greece with the development of societies and the adoption of centralised governance.

Lessons on Implementing Standardisation in Packaging

Implementing Standardisation in the packaging of vessels concerning the ancient Greek Amphoras offers valuable lessons. These lessons are helpful in trade efficiency and record keeping.

Consistency and Uniformity:

Ancient Greek amphoras were manufactured with high uniformity and a good degree of consistency. This standardisation helped the merchants and the consumers understand the quality and quantity of the goods. The ancient Greeks were always conscious of the quality of their productions. So, in modern commerce, the business can easily establish standardised packaging sizes for specific products. It can easily simplify your distribution, inventory management, and consumer expectations. Modern businesses should ensure clear packaging, labelling and design to convey information effectively.

Marking and Identification:

Amphoras were marked with signs, symbols, and inscriptions. They provide essential information about the product's quality and the source of that product. This was a great strategy that the ancient Greeks adopted. So, modern businesses can gain many benefits by employing labelling and bar coding systems for marking the packages. The labelling should contain clear information, including the product name, ingredients, origin, and expiration date. This will help modern companies with quality control and easy traceability, and it will comply with the regulations.

Specialisation for Different Goods:

Ancient Greeks used different types of amphoras for different types of products. It helped in optimising the storage and transportation of goods. Their main focus was producing different types, styles, and designs of those amphoras. So, tailoring the packaging to suit the characteristics of the product and the size of the container will be beneficial for modern companies. It will minimise waste and enhance efficiency and supply chain management. So, in today's world, companies should use specialized containers for different types of products, including solids, liquids, fragile items, and perishables. It will be a good strategy for enhancing commerce and maximizing protection.

Recycling and Sustainability:

Ancient Greeks recycled many amphoras. They were reused for different purposes, which promotes sustainability. So, Modern nesses should also encourage eco-friendly packaging materials that should be recycled. They should also encourage recycling practices that benefit the company and the environment. The business should promote circular economy principles by producing recycling and re-purposing packaging materials.

By applying all these lessons from the ancient Greek Amphoras, modern businesses can enhance transparency, sustainability, and efficiency in packaging practices. It will also lead to more clear and advanced trade and commerce in storing and transporting goods.

Standardisation in packaging is very important for modern businesses, as exemplified by Ancient Greece. It provides transparency, efficiency, and consistency, which benefits both the customers and the businesses. Standardisation streamlines the processes by enhancing the operational efficiency and reducing the risks. Also, standardised packaging helps maintain the product's quality and integrity.

Modern businesses should adopt the standardisation of packaging. It will help in improving competitiveness and improve customer expectations. By adopting effective labelling and marking systems, establishing clear packaging and embracing recycling materials can promote uniformity and sustainability.

Chapter 3

MARKETPLACES AND AGORAS IN ANCIENT GREECE

The agora was more than just a marketplace in ancient Greece. While it is often translated as an open place for assembly or a marketplace, agoras were primarily spaces for selling goods and for meetings and discussions. Moreover, they held significant political and economic importance in the country. This place served people from different occupations for conducting business, holding meetings, and even carrying out religious rituals and prayers.

Furthermore, agoras served as the foundation of democracy, where people openly discussed politics, and men exercised their voting rights in crucial political decisions. Most significantly, agoras were bustling marketplaces where everything from wine to slaves was bought and sold, using barter systems and the Greek currency Drachmae. In this chapter, we will delve into these agoras, their significance in the history of Greek trade, and the valuable lessons they can impart.

Description of Greek Marketplaces (Agoras)

Undoubtedly, Greek cities had various agora or assembly places for their citizens, but the evidence highlights the significance of the Athens agora. For instance, during the 5th and 4th centuries BCE, there were two distinct types of agoras in ancient Greece, as described by Pausanias in the 2nd century CE: the "archaic" and the "Ionic." The "archaic" agoras, like the one in Elis built after 470 BCE, lacked coordinated design, featuring colonnades and buildings that seemed haphazardly arranged, creating an overall sense of disorder. In this section, we will focus on our findings about the Athens agora and its historical importance in the context of ancient Greek trade and economic prosperity.

The Athenian agora was situated just northwest of the famous Acropolis. Moreover, it was conveniently close to the bustling port city of Piraeus. This strategic location placed it right in the heart of one of the busiest and most densely populated city-states in ancient Greece. Before we explore the significance of trade in the Athens Agora, let's take a quick look at what made the Athens Agora unique.

Before the construction of the Athenian agora, this place had a rich history. It dates back to the Neolithic Period when people used simple tools. Later, it transformed into a burial ground. During the Mycenaean civilisation (around 1700-1100 BCE), they established a significant fort on the Acropolis, which overlooked the future agora. However, around 1100 BCE, the Mycenaean Civilisation began to decline, possibly due to climate changes or invasions. Nevertheless, the Mycenaeans left a lasting legacy, featuring prominently in stories like Homer's Iliad and Odyssey, where heroes like Achilles and Odysseus were

Mycenaean Greeks. This historical richness made Athens, particularly the area surrounding the Acropolis, truly remarkable, as celebrated by the writer Hesiod.

During the peak of ancient Greece, the Athenian agora featured renowned buildings such as the Temple of Hephaestus, Agoraios Kolonos, Poikile Stoa, The Stoa of Attalos, and The Prytaneum. These structures played crucial roles as religious, political, economic, and cultural centres for the people of ancient Greece.

Economic Activities Within the Agora

In addition to their political and cultural significance, agoras were primarily bustling marketplaces and trade hubs. People, especially farmers, would gather to buy and sell their needed things. The drachmae, a type of ancient Greek silver coinage, was commonly used for trading within this market.

Nevertheless, this story has a darker aspect because slavery was a significant part of the ancient Greek economy, and people would buy and sell slaves in these markets. However, it's essential to consider that, in the context of that time, domestic slaves were a crucial part of ancient Greek civilisation. Nearly every farmer required one or two domestic slaves for farming purposes. Thus, slaves were a common commodity in the ancient Greek agora.

Besides the slaves, there is a long list of trading goods and services traded in Agora, but the most famous were wine, oil and pottery. Moreover, daily utility goods were also traded in these marketplaces. Fabrics to make clothes and bedding were other common items in the marketplace. These fabrics were colourful and had pretty designs. They also sold ready-made clothes. Women wore long dresses made of linen or wool, and men wore similar outfits but shorter. Sometimes, they wore hats, and women had special undergarments too.

Furthermore, archaeologists have unearthed pottery fragments at ancient agora sites in Athens, Greece. These discoveries include cooking pots, serving bowls, and dinner plates. They also found clay vases, small statues, bottles, and jars for storing honey and other goods. These archaeological finds provide valuable insights into the daily lives and trade practices of ancient Greeks.

It is safe to say that people in ancient Greece incorporated these pottery items into their daily routines, and they were a common sight in the marketplace. Pottery-making held significant importance in ancient Greece, and the artists frequently adorned their creations with meaningful paintings and intricate designs. These pottery pieces not only served practical purposes but also reflected the rich artistic and cultural heritage of the time.

In addition to the items mentioned earlier, the ancient Greek agora was a bustling marketplace where you could find an array of goods. This included fresh fruits and vegetables, culinary delights, exquisite jewellery, and even oil lamps. The agora was truly a one-stop destination for the diverse needs and desires of the people in ancient Greece.

Significance of Greek Marketplaces

With the information mentioned above, it becomes clear that agoras played a vital role in the country's prosperity. They served as the nation's heart, bringing people from different walks of life together on a common platform, facilitating interaction and exchange. Moreover, several aspects highlight the significance of agoras in ancient Greece.

Economic Significance

Agoras served as an essential trade hub, catering to both the local and international needs of ancient Greek traders. This inclusive marketplace allowed everyone, from farmers to shipbuilders, to conveniently acquire necessities in one centralised location. Additionally, agoras showcased the high-quality goods and business acumen of ancient Greek traders to their international counterparts. Acting as intermediaries– traders from different civilisations, one could engage in import and export activities, significantly impacting the country's economy.

Social and Civic Importance

Agoras served not only as bustling marketplaces but also as gathering places for assemblies and social interactions. This was particularly crucial in ancient civilisations/1where men held more prominent positions, and women were often excluded from such marketplaces. In this setting, men conducted business, made significant decisions, and exchanged political ideas with one another. The positive side was that these social gatherings facilitated the exchange of like-minded ideas, fostering innovation and providing opportunities for education and personal growth.

Cultural and Intellectual Centres

Agoras were significant hubs for promoting art, philosophy, and education. These spaces brought together renowned philosophers and artisans to exchange ideas, making them great sources of inspiration for intellectual minds. They were pivotal in fostering creativity and intellectual development in ancient Greece.

Furthermore, through archaeological evidence, we can gain insights into the cultural heritage of ancient Greece from the agora. This understanding highlights how the agora attracted people from other civilisations to engage in cultural exchanges with ancient Greece. It served as a vibrant melting pot of ideas and traditions, contributing to the rich tapestry of Greek culture and its influence on neighbouring civilisations.

Lessons on Establishing Centralized Trade Hubs and Marketplaces

Certainly, in today's world, we have numerous modern trade hubs and marketplaces, including malls and online platforms like Amazon. However, there is always room for improvement, and we can glean valuable lessons from the agora of ancient Greece. These historical marketplaces offer insights that can enhance our understanding of commerce, social interaction, and cultural exchange, even in our modern times.

Creating a Centralised Hub

Certainly, in today's world, we have numerous modern trade hubs and marketplaces, including malls and online platforms like Amazon. However, there is always room for improvement, and we can glean valuable lessons from the agora of ancient Greece. These historical marketplaces offer insights that can enhance our understanding of commerce, social interaction, and cultural exchange, even in our modern times.

Invest in infrastructure

Furthermore, investing in infrastructure development is crucial, considering that agoras were not just functional but also beautifully crafted, representing the rich culture of ancient Greece. This teaches us that the design and interior of your business premises can define your business identity. For example, owners of five-star hotels can decorate different parts of their establishments based on the cultures of different ethnicities, aiming to attract customers from those specific backgrounds. This approach showcases the enduring importance of aesthetics and cultural appeal in business.

Attracting Diverse Traders

Imagine a marketplace where you can find anything, eliminating the need to shop in multiple places. Grocery malls today serve as excellent examples of such marketplaces. Similarly, ancient Greece recognised the value of this strategy and made agoras accessible to everyone. These agoras welcomed people with diverse mindsets, purposes, and ideas, except those with criminal records or women, reflecting the inclusive nature of these ancient marketplaces. This approach echoes the convenience and inclusivity we value in modern shopping centres.

Likewise, modern businesses can create spaces that cater to people from all walks of life. Moreover, even businesses specialising in one product type can employ this strategy by promoting their local culture. For instance, they can offer special teas or drinks from their region's cuisine to attract customers from different cultures and generate more interest. This reflects how embracing diversity and cultural exchange can be a successful approach in contemporary commerce.

Overall, Agoras were the vibrant centres of ancient Greek cities, offering much more than mere marketplaces. They were vital meeting points for discussions, political discourse, and religious rituals. Furthermore, they played a significant role in nurturing cultural exchange, fostering art, philosophy, and education, transcending their role as mere markets. In this context, valuable lessons can be drawn from these agoras, such as selecting strategic locations, investing in infrastructure, attracting diverse traders, facilitating cultural exchange, and promoting tolerance. These lessons provide valuable insights for creating successful modern trade hubs that can serve as dynamic centres of economic, social, and cultural activity.

Chapter 4

ANCIENT GREEK
TRADE ROUTES AND COLONIES

The Greeks made many efforts in the colonisation, which was a very significant historical phenomenon. It has continued for many centuries. All these efforts have established Greek colonies throughout the Mediterranean, the Black Sea, and other areas near it. Many states and cities in Greece have faced overpopulation, which has led to the need for new opportunities. Regarding this, Greeks established trade networks and acquired valuable resources, including timber, metals, and agricultural products.

The cities and other states in Greece have set out many colonies as a way to resolve internal tensions and reduce the risk of civil war. Greeks aimed to spread culture, language, and religion to other religions by forming colonies. The earliest Greek colonies were established in Anatolia, Sicily, and the Black Sea region. The most common examples of the colonies include Byzantium and Miletus.

Overview of Greek Colonisation Efforts

Greeks established their colonies in Southern Italy, including Cyrus and Croton. This Greek spread to different regions, including Spain, North Africa, and Southern France, where Massalia was a prominent colony. Many Greek colonies have focused on the establishment of farming communities and agriculture. Some colonies have become important trade centres due to their strategic locations. These include Massalia and Byzantium.

Additionally, the Greek language, art, and religious practices also spread to different colonies. It also facilitated trade and cultural exchange between the Mediterranean, Black Sea, and other regions. Overall, the Greek colonisation efforts contributed to the expansion of Greek influence, the establishment of Greek states, and the dissemination of Greek culture throughout the ancient world.

Exploration of Greek Trade Routes

The Greek trade routes have been crucial in exchanging culture, goods, and ideas throughout the Mediterranean region. The Aegean Sea was the main hub for Greek trade. It connected numerous cities and states of Greece. The Greeks have also established colonies along the black sea coast. These included Sinope and Byzantium and were available for accessing resources like valuable metals, grains, and fish.

The Greek merchants also engaged in trade with Egypt, especially in Alexandria. It was a major trading hub, and the Egyptian grain was a valuable commodity. Greek colonies have also served as the trade links between the Eastern Mediterranean regions. These colonies include Sicily and Magna Gracia. The colonies were established by Greeks, with access to local resources and trade with the indigenous people. Greeks also exported olive oil, grains, olives, and wine. They were known for maritime trade.

So, they formed different colonies and trade routes, including the Silk Road, which facilitated trade between the Greek colonies and different regions. The trade routes were not only for exchanging goods but also for exchanging ideas and culture. Through these routes, Greek philosophy, language, and art also spread to distant lands. So, the Greek trade routes were instrumental in shaping the ancient world. It has posted the connection between the Greek civilisation and other regions.

Importance of the Lucrative Black Sea in Ancient Greece

The Black Sea trade route had a very significant importance in the ancient Greek world. This region was rich in numerous valuable resources, including timber, metals, fish, and grains. The cities and states of Greece relied on all these resources to support their population and economic growth. Moreover, the fertile land around the Black Sea was also a main source of grain provision for the population of Athens, the largest state of Greece at that time. This Black Sea trade route allowed the traders of Greece to engage in profitable commerce.

In exchange for the Black Sea commodities, they have exported numerous goods, including wine, olive oil, and pottery. This trade has increased the economy of Greek states and has helped them in accumulating wealth. The lucrative trade opportunities in the Black Sea also motivated the creators to establish numerous colonies along the coast. The notable colonies of the Greeks include Sinope and Byzantium. These colonies have been a great trading post that has contributed to the expansion of Greek cultural influence in that area.

Also, the location of the Black Sea has made it strategically important. This trade route allowed the Greek merchants to influence the neighbouring regions and strengthen their geopolitical standing. Moreover, the Black Sea trade route was sometimes a source of competition between regions. Battles for controlling the Black Sea trade route were very common in ancient history. So, the Black Sea trade route was essential for ancient Greece. Because it provided vital resources for economic growth, promoting culture had a very important role in the prosperity of Greek civilisation.

Colonies of Magna Gracia in Southern Italy

Magna Gracia was a region located in Southern Italy. Here, many colonies were established during the ancient period. All these colonies have played a very important role in the diffusion of Greek culture. They also helped in the development of Western Mediterranean culture. The Greeks established many colonies in the strategically located area of Magna Gracia.

Syracuse:

The most prominent colony in Magna Gracia was Syracuse. It was located on the eastern coast of Sicily and was one of the most powerful Greek colonies in that region. It was known for its diverse culture, Greek philosophy contributions, and military strength. This colony had a great name in education, philosophy, and culture.

Tarentum:

Another notable colony that the Greek States created was Tarentum. It was situated on the Italian Peninsula. It was a main trading hub known for its wealth and strong navy. This colony played a very important role in the trade networks with other Greek colonies and Italian people.

Rhegium:

Rhegium is another Colony created by the Greek States. This colony was located at the southern tip of Italy. It is situated between the passageway of Italy and Sicily, which makes it strategically important for naval control and trade. Many commodities have been traded through this route. Due to this, the Greeks have benefited greatly in money and culture.

The colonies of Magna Gracia were the centres of Greek culture, including philosophy, literature, architecture, and art. They have helped in spreading the Greek influence throughout the Western Mediterranean region. Moreover, all these colonies were agriculturally productive as well. They were engaged in trade with other colonies, Phoenicians, and the Italian people contributed to the region's economic prosperity.

Provision of Access to Valuable Resources by Greek Colonies

The Greek trade routes and colonies have significantly provided access to variable resources and expanded opportunities. Greek colonies were abundant in agricultural resources. They produced crops like grapes, olives, barley and wheat, essential for growing populations. Some colonies in Greece also provided excess valuable metals like copper, iron, and silver. It was very crucial for military equipment and coinage. The colonies constructed in the forest regions provided timber and other materials supporting construction and shipbuilding work.

Lessons on Identifying and Establishing Trade Routes and Colonies

Adopting the strategies of ancient Greece for establishing the trade routes and colonies offers numerous valuable lessons in Geopolitics, economics, and geography.

Understanding Geographic Advantage:

Geography has played a very important role in the Greek trade and colonisation in ancient times. The Greeks have recognised the importance of strategic locations, including natural harbours, islands, and coastal regions. They recognised the significance of the trade routes and colonies for defence and trade. So, when identifying the trade routes and potential colony sites, they always considered geography. Modern businesses should also find findings setups for their companies with access to transportation resources and defensible positions. It can provide a big advantage to modern businesses in the today's world. They should consider geography a key factor in locating their colony sites or trade routes.

Economic Factors:

Many economic factors were the driving force behind the Greek colonisation efforts. Overpopulation, economic opportunities, and resource scarcity motivated the Greeks to establish different colonies and trade routes. Similarly, modern businesses can easily analyse the economic factors in their area while considering the trade routes or colonies.

They should assess the market demand, availability of resources, and the potential for economic growth. This will be very helpful for them in increasing their business potential in the market.

Political and Cultural Influence:

The Greek colonies were not only beneficial for economic gains, but they were also significant in spreading Greek culture, ideas, and language. So, in ancient times, cultural and political considerations played a very important role in establishing Greek colonies and trade routes. About the Greek people, modern businesses can easily recognise the potential for political and cultural influence.

When they are going to establish the trade routes, they must consider their presence in the market as well as what effect their presence will have on the local culture and politics. By assessing all these factors, modern businesses can easily excel and have a great cultural and political influence on other societies.

NavAl Strength:

Greeks were very well-known sailors, and their naval strength was notable in securing the trade routes and protecting colonies. The naval power was important for projecting influence and for defence. It has helped the Greeks in promoting trade. So modern businesses should also prioritise the development of naval capabilities in the colonies and trade routes near water. They should know that the prolonged naval presence will safeguard their products and business. So, the navel strength must be optimised by modern businesses to secure their colonies.

Understanding the significance of the colonies, the trade routes, and the Greek Civilisation provides numerous lessons for modern-day businesses. The colonies and trade routes were the key drivers of economic prosperity, cultural exchange, and geopolitical influence. Modern businesses should assess the Geographic advantages of the cultural sensitivity diversification of the trade networks and develop a strategic plan for assessing the potential benefits for their business.

By identifying the trade routes, businesses can easily gain a competitive edge in the market. By strategically establishing the targeting markets and trade routes, businesses can easily expand their customer base and increase revenue. They can get many lessons from the ancient Greece. As Greeks have leveraged the colonies and trade routes for prosperity, modern businesses should also benefit from the global perspective and strategic expansion.

Chapter 5

COINAGE AND CURRENCY SYSTEM OF ANCIENT GREECE

Ancient Greek coins represent adaptability, innovation, and a significant step toward technological advancement. Greek coins have a rich history and valuable lessons for traders. The implementation of standardised coinage systems in ancient Greece facilitated extensive trade and brought efficiency to transactions. This represented a technological leap forward, as Greek traders, after centuries of bartering, recognised the need for coinage in both local and international trade.

The evolution of the ancient Greek currency system was a multi-stage process marked by innovation. It extended beyond the design and symbols on the coins, as they made concerted efforts to ensure the intrinsic value for import and export. After all, without the inherent value of money, it becomes meaningless. Ancient Greece's journey through these stages of coin development and innovation reflects their commitment to enhancing trade and the efficiency of their economic system.

In this chapter, we will delve into the innovative history of coinage and the currency system of ancient Greek traders. We will also explore the valuable lessons modern traders can glean from this historical context, shedding light on the enduring principles that continue to shape economies and trade today.

Historical Development of Greek Coinage Systems

The coins of ancient Greece indeed hold the esteemed status of being the oldest in existence, offering compelling evidence of a remarkable civilisation. As previously mentioned, Greek coinage highlights the adaptability and forward-thinking nature of Greek traders. While many other civilisations were still tethered to barter or weight measurements for trade, ancient Greece swiftly embraced and incorporated coins into their commercial practices.

To delve further into the fascinating history of the coinage system, let's break the chapter into the following categories, providing a structured and insightful exploration of this important aspect of ancient Greek culture and its impact on trade.

Pre-coinage barter and early forms of currency

Before the 6th century B.C.E., ancient Greek traders heavily relied on the barter system. However, the barter system had its limitations. It didn't guarantee equal value exchanges between two parties, potentially leading to inequality if one party possessed more valuable goods. For instance, consider a situation where a person exchanged a sheep for just one sack of grain. This demonstrated the need for a more reliable currency system.

However, the emergence of a currency system wasn't an instant occurrence; it developed over several years. Before the introduction of coins, traders began using metal bars made from materials like bronze or copper for exchanging goods, but this method required mutual agreement between both parties. Later, they transitioned to using metal rods or obelo, approximately 1.5 meters long, for the same purpose.

The most interesting aspect is that these metal rods and obeloi were physically graspable, and in Greek, the word "drattomai" translates to "grasp." Hence, it's plausible that the name for the silver coins, "drachmae," derived from these very words. The utilisation of these bars and obeloi laid the fundamental groundwork for the coinage system that ancient Greek traders would eventually embrace and refine.

Emergence of the first standardised coins

The seventh century B.C. marked a significant turning point when true money, not merely tools for measuring or valuing goods, was introduced. It's crucial to emphasise that we're discussing the emergence of genuine currency here. Ancient Greece was at the forefront of inventing and employing money internationally. Although there's some debate about the birthplace of the first coin, most evidence points to Lydia, which is in present-day Turkey. According to the historian Herodotus, these coins were minted by order of King Croesus of Lydia and are known as "Creseids." King Croesus acquired gold from the Pactolus River in Lydia during that period, marking a significant development in the history of currency.

During that era, most ancient Greek traders embraced the coinage system with open arms as it became necessary. It was particularly vital for paying soldiers who couldn't feasibly carry heavy barter items from one place to another. Since soldiers came from various regions, the use of coins spread with them.

What's intriguing is that in contrast to the modern world with advanced security systems, there were two methods of safeguarding coins during that time. The wealthy often buried their wealth underground, as we can find evidence in the archaeological record. In contrast, the less fortunate had to carry their money in their mouths to prevent theft, given that they possessed only small amounts.

Furthermore, it was during this period that every city started to produce its currency instead of relying on coins made in Lydia. As a result, much of the system still relied on the barter system.

Additionally, just like in our world, the value of coins could change from one era to another in ancient Greece. For instance, the drachmae were made of silver and were worth about a week's work for most people in the 5th century.

Evolution of coin designs and materials

Minting coins

The coins were intentionally lighter than their pure metal counterparts, allowing the coin miners to profit from the process. The minting was a manual procedure, with most coins featuring stamps on both sides, although some had designs on just one side. Coin makers employed specialised tools, utilising a die for one side and a punch for the other. The process began with cutting appropriately sized pieces from metal cylinders. An artist then carved the design into hard bronze or iron dies. Placing a blank piece of gold, silver, or electrum (a gold-silver alloy) in between, they used a hammer to create the designs on both sides.

Material

The majority of coins were crafted from valuable metals like gold and silver. Silver, exemplified by the drachmae, was the most common material for coins, while gold coins were quite rare and mainly owned by the wealthy elite. In addition to these precious metals, electrum, a naturally occurring blend of silver and gold, was also employed in coinage.

On the other hand, low-value metals such as copper, bronze, and lead found their place in coin production. However, these metals weren't well-suited for complex coin designs.

Design and weight

Various cities had distinct coins, each varying in weight and design to reflect their identity. Furthermore, these coins had different values, like the obol (six of which equalled one drachma) or the double octadrachm. Yet, Athens stood out by adhering to a consistent weight system known as the "Attic" standard. An example is the Drachmae, which contained 4.3 grams of pure silver. The Athenian coins' exceptional purity made them highly sought after in trade, gaining widespread acceptance among the people.

Athens not only established the "Attic" standard but also pioneered the practice of adding symbols to their coins. As their coinage gained renown, other cities' coin makers began to emulate their approach. While most of these coins were circular, it was the unique stamps and sizes that set them apart. It's intriguing to note that the earliest coins featured geometric shapes, often a quartered square, on their reverse sides.

Furthermore, the stamps on these coins typically bore symbols associated with Greek gods and characters from Greek mythology. Rulers also employed coins to imprint their likenesses, forging a connection with gods and heroes to demonstrate their authority. Moreover, nearly every city adopted a unique emblem to symbolise its identity, such as Athens with its wise owls or Corinth featuring Pegasus, the legendary winged horse.

Overall, coins marked a significant development in ancient Greece, relevant to all. While other artefacts like pottery may find their place in museums, coins possess a unique value due to their composition of precious metals and economic significance, a legacy handed down to us by ancient Greece.

Influence of Greek coinage on Greek trade

No doubt, ancient Greek coins hold both artistic and cultural significance, offering insights into their civilisations and showcasing remarkable artistic craftsmanship. While they serve as valuable historical artefacts, their primary function remains economic and monetary, a testament to the dual role of these remarkable pieces of history.

As we know, Greeks were renowned as great traders, exchanging a wide array of goods with other civilisations. Their pioneering use of coins played a pivotal role in enhancing trade. Coins offered a standardised means to estimate the value of goods, making transactions more efficient and reliable.

Civilisations/1like the Phoenicians, who desired precious metals in the form of coins, engaged in trade with the Greeks. Particularly, Athens coins, crafted from high-quality metals with accurate silver and, at times, gold content, earned a reputation for their reliability. This contributed to the unstoppable growth of trade with the Greek civilisation, especially after establishing a standardised coinage system around the seventh century. In summary, using coins significantly facilitated and prospered trade for ancient Greek traders.

Lessons on Developing a Reliable Currency System

The invention of currency in ancient Greece is proof of their innovation and adaptability in the face of advancements and the need for money. They understood the significance of currency, which was accepted by other civilisations, so they tried to keep the value of money high in the realm of international trade. The following are the lessons we can learn from them.

The need for standardised and trusted currency

Many countries in today's world are facing the devaluation of their currency for various reasons. Despite being one of the oldest civilisations, Ancient Greece understood the importance of a high-value currency system. As we know, Athens' coins had value and were in demand from other civilisations, which helped them prosper in their trades. Even civilisations that didn't embrace a monetary system traded with them due to the intrinsic value of their metal coins. Modern businesses should take a page from this guidebook from Ancient Greece, emphasising the necessity of well-crafted, reliable, and standardised financial systems.

Role of authorities in coinage regulation

The first money was invented on the order of King Croesus of Lydia, signifying the importance of authority. As an example, it highlights the responsibility of authorities to make decisions while understanding market demands, much like what the King of Lydia did. As we know, businesses can be led by strong and determined leaders. In this way, businesses should invest in themselves and their employees to become leaders and guide others in the right direction.

Challenges and pitfalls in currency development

One of the most important lessons the ancient Greek currency system can teach modern businesses is handling unsuccessful products. As we discussed earlier, ancient Greece attempted to produce coins from cheaper metals like bronze or lead, which proved to be a pitfall as these were not accepted by traders from other civilisations/1who dealt in precious metals. Instead of completely discarding them, ancient Greece learned from this experience and used these coins for very small transactions at a local level. Modern businesses can also understand the need to adapt and learn from such experiences.

While ancient Greece pioneered the coinage system, they employed it for various purposes, with international trade being a major driver. It's worth noting that the barter system was still prevalent when the coinage system was introduced in Greece, leaving a profound impact on Greek trade and other civilisations. These coins, adorned with artistic designs and mythological symbols, stand as archaeological treasures. Moreover, coins offered a convenient means for individuals to save money for the future. As a result, the ancient Greek coinage system imparts enduring lessons for modern economies, highlighting the importance of standardised, trusted currencies in promoting trade, trust, and economic growth.

Chapter 6

MIDDLEMEN AND INTERMEDIARIES

Trade has long relied on intermediaries to link buyers and sellers across distances, enabling exchange and commerce. These middlemen play an instrumental role in expanding market access and forging connections. Ancient Greek merchants exemplified such intermediaries, strategically situated along Mediterranean trade routes where they became enterprising middlemen catalysing vibrant international exchange networks. The successes of these Greek traders highlight timeless lessons for modern intermediaries operating globally.

This chapter examines the historical role of Greek merchants as intermediaries in antiquity and the specific ways Greek merchants expanded market reach, reduced transaction costs, and built relationships between trading partners. Their successes highlight the continued importance of intermediaries in today's global economy. The in-depth analysis in this chapter will provide key takeaways for modern intermediaries.

Vital Role of Greek Merchants as Intermediaries

In ancient times, Greek merchants were the crucial intermediaries that enabled extensive maritime commerce and trade between distant regions surrounding the Mediterranean Sea. They operated complex logistical networks that maximised trade opportunities for Greek city-states and colonies. They imported key commodities to Greece, like grain, metals, papyrus, fabrics, and other goods from Egypt, Cyprus, the Levant, Sicily, and beyond. The merchants also exported Greek commodities prized by foreigners, such as olive oil, wine, silver, and pottery. This vibrant two-way trade supported the prosperity of Greek city-states and colonies.

Prominent Greek maritime powers like Athens and Corinth derived great wealth from the activities of their overseas merchants as they connected producers and consumers across continents. For instance, the Greek colony of Naukratis in Egypt emerged as a thriving hub for Greek merchants and their trade partners from diverse origins.

The Greek merchants specialised in facilitating transactions between parties in different parts of the Mediterranean trade network. By bridging vast geographic and cultural divides, they enabled economic growth for all participants involved in the long-distance trade.

Greek Mercantile Practices and Strategies

To succeed as trusted intermediaries, Greek merchants employed a variety of innovative business strategies and practices. Their specialised knowledge, experience, and relationships facilitated complex transactions between distant regions.

One key mercenary strategy was establishing trade colonies and settlements across Mediterranean ports to extend their trade influence and networks. For example, Greek traders secured special legal privileges and trading rights in key emporia like Naukratis in Egypt. This provided a legal and physical base for conducting regional trade.

The merchants also leveraged relationships with political elites to gain preferential trade terms and status. By becoming the designated suppliers to royal or aristocratic courts, they cemented their position in high-value trade networks.

Financial practices like maritime insurance, currency exchange, and credit enabled Greek traders to operate across diverse regions with varying forms of money and laws. Their financing services reduced risks and costs for trade clients.

The merchants invested heavily in building their reputations for honesty and reliability over repeated interactions. This attachment to reputation encouraged fair dealing, as cheating would be punished by community exclusion.

Network Structure and Operations

The structure and operations of Greek trading networks maximised efficiency. At its peak, the trade interconnected hundreds of colonies, ports, and production centres across thousands of miles. Regional commodities flowed through smaller provincial ports towards larger urban hubs, offering both consumption markets and redistribution networks. Key distribution centres, like the free port of Delos, enabled goods to be sold across multiple trade lines.

The extensive connectivity between producers, merchants, and ports allowed the networks to function smoothly and respond to disruptions. Alternative routes ensured continuity if a link was disrupted by piracy or weather. Cooperation and divisions of labour between merchants specialised in different routes, commodities, and regions optimised operations. Partnerships allowed merchants to spread risks and upscale.

Decline of Greek Maritime Trade and its impacts on middlemen businesses

Greek merchants dominated Mediterranean trade for centuries, but their intermediary networks eventually declined as rising Western powers like Rome challenged their control over maritime commerce. Political infighting between Greek city-states and growing piracy disrupted trade networks. By the 2nd century BC, Roman annexation of Greek territories ended Greek dominance over Mediterranean trade, though Greek maritime knowledge continued shaping regional commerce.

Here is a summary of the impacts of the decline of maritime trade on intermediary traders and their businesses.

Reduced business and profits:

With less maritime trade volume, traders would have fewer opportunities to buy and sell goods, reducing their overall business and profits. Lack of trade means a lack of income for traders.

Increased overhead costs:

Fixed costs like warehousing, transportation, and labour would be spread over lower sales, increasing overhead costs per unit handled. This cuts into traders' margins.

Narrower product range:

Traders would have access to fewer exotic goods with reduced global trade. This limits the diversity of their inventory and product offerings.

Shift to local/regional trade:

Traders would rely more on local and regional exchange networks versus international maritime routes. This localised trading restricts market reach and margins.

Business contraction and closures:

A persistent decline in maritime trade would force many intermediary businesses to downsize operations or even close down altogether.

Changing business models and diversification:

Traders might adapt by shifting to alternative goods with local demand or diversifying into other services. However, new ventures may have lower profit potential.

Economic decline of trader hubs:

Major port cities and trade hubs would see an economic decline as trader enterprises flounder, impacting local economies.

In summary, the contraction of maritime trade networks dealt a huge economic blow to intermediary trader enterprises dependent on international exchange to thrive and survive.

Continued Relevance in the Modern Era

Despite their eventual decline, Greek merchants show intermediaries remain essential for trade across distances. Each producer needs help managing global operations. Intermediaries bridge divides between producers and consumers separated by geography and culture, benefiting both sides. While technology transforms trade, personal relationships and expertise stay pivotal for transactions. Global intermediaries should adopt Greek merchants' relationship-building and risk-minimizing practices. This underscores why Greek intermediaries provide vital lessons for expanding connectivity today.

Implications for Modern Intermediaries

The impressive accomplishments of ancient Greek merchant intermediaries continue to provide crucial lessons for intermediaries operating globally today. Modern intermediaries must recognise the importance of establishing credibility and effective communication with all parties involved in a transaction. While modern technology assists global trade, personal relationships remain pivotal.

By specialising in understanding both buyer and seller needs, modern intermediaries can strategically bridge cultural divides and geographic distances. This enables participants to benefit from trade based on their strengths and needs.

Legacy and Lessons for modern businesses

Fundamental lessons that remain relevant today emerged from the Greeks' business dealings. They prioritised establishing personal ties and trust between traders through cooperation over multiple interactions. Maintaining strong community standing and credibility was valued. Adapting to regional differences in language, customs, and practices was key in new territories. Their agility in developing trade alternatives allowed them to overcome disruptions. Modern intermediaries can draw on several timeless lessons from the Greek traders:

- Modern companies ought to comprehend the cost of intermediaries. It is important to recognise the pivotal function of Intermediaries. They need to respect that intermediaries provide admission to new markets, clients, and opportunities. Recognising the cost of intermediaries permits agencies to leverage their expertise, networks, and sources for a boom.
- Taking time to develop rapport, familiarity and trust between trade partners is essential in building resilient long-term networks.
- Consistency in delivering value and maintaining a strong reputation breeds credibility with customers.
- Leveraging intermediaries for market expansion. In other words, collaborate with e-commerce systems, global distributors, and different intermediaries to reach new markets and patron segments. Strategic partnerships with intermediaries enable businesses to reach a worldwide target market beyond geographical barriers.
- Reduce transaction costs and employ cost-reduction strategies
- Implement price-saving measures such as outsourcing logistics, leveraging economies of scale, and using era to streamline operations. Efficient fee management no longer complements profitability but also guarantees competitiveness in the international market.
- Businesses must invest in cultural fluency - understanding the languages, norms and customs to penetrate new markets.
- Leveraging collaborative specialisation and partnerships minimises risks and maximises collective strengths.
- Nimbleness, versatility and diversification allow businesses to adapt to changing conditions and sustain operations through challenges.

By following such practices, modern intermediaries can also construct resilient, ethical trade networks that create value for all participants through strong relationships.

In summary, Greek maritime merchants performed the essential functions of trusted intermediaries in antiquity to spur Mediterranean trade networks and prosperity. Their success holds crucial lessons for intermediaries aiming to create value today by bridging distances, minimising costs, and forging enduring global human relationships between producers and consumers. Intermediary practices must evolve with technology but remain rooted in cultivating personal bonds and reputation. By analysing the achievements of Greek merchants, modern intermediaries can refine their strategies to build ethical, resilient, and productive global trade ecosystems. Throughout ancient history, Greek traders leveraged their extensive connections, local knowledge, and reputation for fairness to facilitate complex deals and transport goods safely across vast seas and territories. They exemplified how intermediaries generate trust and enable commerce between disconnected groups through relationship-building and reliable services. Their accomplishments provide timeless insights for intermediaries seeking to enable equitable, transparent, and efficient exchange in today's globalised economy.

Chapter 7

FAIR TRADE PRACTICES
IN ANCIENT GREECE

Ancient Greece may not have had fertile land like other prominent civilisations like the Indus, Egypt, and the Phoenicians, but what set them apart and made them powerful? The answer lies in their remarkable trading skills. Without the principles of fair trade and a strong code of business ethics, it's impossible to build an empire like the Greeks did in the past. Greek traders were guided by these principles – honesty, trustworthiness, and keeping their word. They understood the value of their goods, and this fairness ensured successful deals and promised enduring partnerships that were the cornerstone of their trade success.

Similarly, the code of business ethics has found its place in today's enterprises. Businesses educate employees on ethics using formal codes of conduct, rules, and training programs. Additionally, ethical behaviour is often demonstrated informally through the actions of company leaders and the corporate culture. However, there are many lessons we can draw from the fair trade practices of ancient Greek traders. In this chapter, we will explore Greek standardisation of measurements, trade incentives and protection, and their interactions with other traders, all depicting their commitment to fair trade practices.

The Greek Code of Fair Trade Practices

From crafting amphoras to trading with ancient Egypt, ancient Greece has left us with numerous examples of fair-trade practices. Among these, their ability to navigate tough times, especially with government involvement, stands out prominently. A prime instance of this government intervention occurred around 470 BCE during a drought when Athens needed to feed its large population. They took charge by controlling and buying wheat through a designated buyer, ensuring a reliable and equitable supply of this vital commodity. This demonstrates how ancient Greece's fair-trade practices extended even into managing challenges and ensuring the well-being of its people.

To achieve this goal, they established strict rules for grain trading, preventing people from making excessive profits and ensuring that grain was sold at fair prices. The government even went to the extent of imposing the death penalty for those who broke these rules. This stern approach emphasised their commitment to fair trade practices and the importance of providing essential goods like grain to the population at reasonable rates.

Additionally, they appointed officials to monitor the quality of goods sold in the markets. At the same time, prices were regulated to account for variations in quality. Specific supervisors, known as sitophylakes or agoranomoi, were tasked with ensuring that prices and quantities were accurate. This dual approach maintained fairness in trade and guaranteed that consumers received products of the expected quality, reinforcing trust in the marketplace.

By implementing these measures, Ancient Greece showcased a strong dedication to fair trade practices, particularly for vital commodities like grain. This approach had a twofold purpose: to establish economic stability and to guarantee food security for its people. Moreover, it promoted ethical behaviour in trade dealings, emphasising the enduring importance of honesty and fairness in commerce, lessons that resonate even in modern times.

Providing Security for Fair Maritime Loans

Maritime loans were a vital tool for boosting the economy and ensuring trade incentives and protection, and ancient Greece excelled in this practice. These loans were offered to small and large merchants without discrimination based on race or ethnicity. They provided traders with the means to pay for their cargo, and Greek traders and shipbuilders offered security to minimise risks. For instance, if a ship failed to reach its destination safely, the borrower only had to repay the loan if it arrived safely; otherwise, no payment was required. This maritime loan system exemplified fair trade practices and moral values.

However, there was a drawback to this system as the interest rates were relatively high, ranging from 12% to 30%. These rates were set to mitigate the risks involved, as cargo dealers faced substantial risks in safeguarding their money and protecting the goods of fellow traders. Despite the higher interest rates, this approach highlighted the Greeks' commitment to fair trade practices and their dedication to ensuring the security of trade transactions.

HonoUring Contracts and Agreements

Honouring contracts was another significant aspect of fair trade practice in ancient Greece. They highly valued their trade and recognised the importance of setting clear standards to maximise profits. Ancient Greek society was not hesitant to establish explicit and concise rules for Greek traders. These rules ensured that both parties involved in a trade would fulfil their contract and agreements with each other, emphasising the importance of trust and reliability in their business relationships.

Numerous associations and institutions played a role in these trade practices. Alongside these, protective measures and taxes were enforced. For instance, in Athens, citizens who borrowed money for grain but didn't bring the grain to Piraeus, the main port, had to pay taxes as a penalty. Similarly, merchants who arrived with cargo but didn't unload a specific portion of their goods faced taxes, too. These measures aimed to guarantee the availability of grain, essential for feeding the city's sizable population, and to ensure that traders conducted their business fairly and reliably in Athens.

Much like Athens, other trading hubs, including Thasos, also enticed traders with incentives. Thasos, renowned for its top-quality wine, probably offered advantages like favourable trade terms or support to attract traders and bolster wine commerce. This strategy aimed to make Thasos an attractive and profitable destination for traders, showcasing the importance of fair trade practices in fostering prosperous business relationships.

Moreover, private banks held a pivotal role in Ancient Greece by aiding in currency exchange and offering a secure haven for individuals to safeguard their money. These banks simplified financial management for both traders and individuals.

Overall, these measures aimed to attract more consumers and ensured fair practices among traders. The knowledge that financial institutions monitored transactions served as an additional incentive for traders to adhere to fair trade practices, fostering trust and ethical conduct in the marketplace.

Lessons from Ancient Greece

With the examples mentioned above, we can understand the fair trade practices and strict rules of ancient Greece that they applied. The following are the lessons we can learn from them.

/1A. Upholding Ethical Business Practices

There is a long-lasting value in the honesty and integrity practised by ancient Greece. For example, their consistent use of standardised measurements enabled their dealers to demonstrate fairness in assessing the value of goods and offering appropriate prices for products from other traders without dishonesty. This fair and transparent practice can serve as a valuable lesson for modern businesses, encouraging them to establish fairness in their dealings with dealers and showcasing their commitment to ethical business values rather than relying solely on advertisements on T.V. screens.

/1B. Building Credibility

Credibility is pivotal in long-term business success; government involvement in grain trade is an excellent example. Greek traders were confident in their investments in the grain trade because they knew the government safeguarded their goods and money, eliminating their worries. Similarly, buyers believed they were acquiring high-quality grain at a fair price, thanks to the government's credibility. For modern businesses, government involvement may not always be necessary to establish credibility. However, implementing strict rules and regulations to ensure the quality of your products at reasonable prices can instil trust in your business, ultimately leading to lasting success.

/1C. Fostering Trust with Trade Partners

Ancient Greece imparts valuable lessons on how to build trust with your trading partners. From honesty and keeping promises to fairness and transparency, these principles are crucial. The ancient Greeks' dedication to these values in their fair trade practices laid a solid foundation for trust, a vital principle in modern business. By adopting these values as part of your business code of ethics, companies can thrive and cultivate long-lasting partnerships.

Ancient Greece demonstrates that offering business security and protection to uphold fair trade practices is not modern. Instead, businesses can revitalise their moral codes with the lessons these ancient traders have left behind and thrive, just as they did to become successful and influential traders in their era.

Chapter 8

CULTURAL EXCHANGE

Business is about building relationships; nothing builds a stronger relationship than having a common understanding of the culture. Cultural exchange has existed as long as trade has. Greek culture can set the tone for business in other cultures. The ancient Greeks were known to be very early adopters of the latest developments in science, architecture, politics, philosophy, and language. Trade was a major factor behind this, as well as colonisation. It allowed for new experiences with different cultures. They traded with other cultures to learn new things and grow their influence.

The Greeks Had a Rich Culture That Blended Artistic Beauty and Natural Elements

The Greeks were known for their love of the arts, especially drama and poetry. They also excelled at architecture, sculpture, and painting. The Greeks' artistic style was influenced by their surroundings: they lived on the coast of the Mediterranean Sea in an area with many different climates and landscapes. The Greeks combined these elements into their art to create beautiful works reflecting the natural world.

Their architecture was based on symmetry, with columns supporting a flat roof. The Greeks used columns to create a sense of grandeur in their buildings. The Greeks also designed their buildings to use natural light as much as possible, creating open and airy spaces.

For example, they would carve marble into statues so realistic that they looked like people who were alive. They did this by studying anatomy and drawing from live models. The Greeks also created beautiful mosaics of colourful stones set into mortar or plaster.

Construction of Parthenon

The Parthenon, for example, was built in the 5th century B.C. and designed to look like Athena's helmet. Doric columns made from marble from Mount Pentelikon surrounded the temple. The columns were topped with capitals resembling plants and animals—a symbol of life on Earth. The roof was made of tiles that imitated leaves on trees, also part of nature.

Artists such as Phidias also incorporated nature into their work. His sculptures depicted gods who embodied certain virtues or aspects of nature, such as Aphrodite (love) or Dionysus (wine).

Ancient Greek Culture Was Critical To Alexandria In Egypt - A World Centre For Learning And Knowledge

The city was founded by Alexander the Great in 332 BCE. It was built on the Egyptian coast of the Mediterranean Sea to be a trade centre between Europe and Asia.

The focus was on mathematics, astronomy, medicine, and other subjects. Many scientists came to study at the Museum and Library of Alexandria. The library contained many books written by famous Greek authors like Aristotle, Euclid, and Archimedes. It also included hundreds of scrolls copied from ancient texts that have been lost forever.

Alexandria quickly became famous for being a home for scholars interested in studying the Greek language, literature, philosophy, music, mathematics, medicine, and science. Many great thinkers lived in Alexandria, including Euclid (who wrote Elements) and Eratosthenes (who calculated Earth's circumference).

For Example, Ptolemy I Soter was a great leader who made many important decisions during his reign over Egypt:

- He started building the Library of Alexandria, which became one of the greatest libraries in history.

- He also started building a museum for natural science research; he organised Egypt into provinces ruled by governors.

- He wrote books about agriculture.

- He established a new calendar based on observations of stars and planets (including Earth's moon).

Ancient Greek Traders Also Introduced New Types of Foods

Ancient Greek traders also introduced new types of foods to other peoples, such as figs, olives, and grapes. These fruits were popular in Greece because they grew well in the climate there. The ancient Greeks had already established trade routes that brought them into contact with many different cultures, and they used these trade routes to share their knowledge about growing grapes and olives with other countries. It helped spread the knowledge of how to grow these foods worldwide, making them more common throughout history than they might otherwise have been.

The Macedonian Ruler Encouraged His Soldiers to Marry Persian Women and Mix with Their Cultures

The Macedonians were a tribe of Indo-European people who crossed into Europe from the Near East. Alexander led them in an invasion of Asia Minor (modern Turkey) in 334 B.C.E. Alexander had already conquered much of Persia by 331 B.C.E. and continued expanding his empire.

Although he did not allow his soldiers to marry native women, Alexander did encourage them to marry Persian women who were often young and beautiful and mixed with their cultures. He believed that doing so would help him create a multicultural society that would be more stable than one based on pure ethnic divisions alone."

It was a smart move on Alexander's part because it helped him create a bond between his men and their new homeland. It also helped to build up the population of Macedonia's cities and towns, which was important for the economy. The soldiers who brought their wives back home to Macedonia had children who grew up there and eventually became an essential part of Macedonian society.

The Macedonian ruler Himself Wed Persian Women

When Alexander the Great conquered Persia in ancient times, he married two Persian women. Alexander was very interested in the cultures of the places he conquered and wanted to understand them better by marrying into them. He married Roxana, daughter of Oxyartes, a Bactrian nobleman. He also married Stateira II, the daughter of Darius III and Parysatis.

First-Hand Experience - Greek Colonies Export Their Culture, And Eventually, The Cultures Merge

Colonisation is when a group establishes a settlement in another area and brings their traditions and customs. The Greeks colonised many regions around the Mediterranean Sea, including Sicily, Southern Italy, North Africa, Spain, France, and Turkey. These colonists brought their language, art, architecture, literature, and other aspects of Greek culture with them. They also took over native religions and replaced them with Hellenistic forms of worship, such as polytheism (the belief in many gods).

When they travelled to new lands, they took their culture with them. One of the ways this happened was through trade. The Greeks traded with other cultures they met in their travels, and sometimes, these cultures would even adopt Greek customs and beliefs. This is why we have so many Greek-influenced cultures today—the Greeks exported their culture all over Europe and Asia!

The influence that Greeks had on these regions is still seen today in many aspects of modern life, from architecture to food preparation methods, which are still used by people living in these areas today despite having been conquered by other groups after they fell under Roman rule."

In addition to trading with each other, the Greeks also traded with different cultures around the Mediterranean Sea region. They traded with Egyptians for wheat and enslaved people and Phoenicians for textiles, wood, glass, metals, incense, papyrus, and carved ivory. These goods were brought back to Greece and sold at marketplaces like Athens' Agora or Corinth's Isthmus Marketplace.

For Example:

In the Archaic period, the Greeks began to trade with other cultures in their region. The first major trade partners were the cultures of southern Italy and Sicily. They are right next to Greece and had close cultural and economic ties to several Greek states. From there, the Greeks expanded and started trading with the people in Egypt, Carthage, Ethiopia, and the Arabian Peninsula.

The pottery produced by these cultures was also very different from what was made in Greece—it was not as refined or complex. It allowed for an exchange of ideas between these groups and helped them develop their unique styles of pottery that would eventually become some of their most famous exports.

The most important type of pottery that they exported was black-figure pottery from Athens. It was made by firing clay pieces on a kiln at temperatures above 1,200 degrees Fahrenheit (648 Celsius). It created a black glaze on top of a white background for decoration. This pottery was used as storage containers called amphoras or kraters (large bowls used for mixing wine) and decorative vases used for display purposes such as funerals or wedding ceremonies. These examples have been found all over Europe, Africa and Asia, which shows how widespread this type of pottery is! The Greeks traded their pottery for metals like gold and silver, spices, and food.

Hospitality is the Key to Cultural Exchange

The ancient Greeks were so dedicated to cultural exchange that they had a word for it: xenia. It's Greek for "guest-friendship," meaning that strangers could enter your home as guests and become your friends—and vice versa.

It was a radical idea at the time because it meant that citizens would have to welcome foreign visitors who might not share their language or culture openly but would receive hospitality nonetheless.

The Greeks believed this was the best way to cultivate meaningful relationships with people from other cultures. So, they ensured everyone knew it was necessary by enshrining xenia in their laws and customs—and even making it illegal not to participate!

The Importance of Cultural Exchange in Business

When you think about the ancient world, you probably think of things like the pyramids and the Roman Empire. But you probably don't associate it with business practices. However, there is much more to learn from these ancient cultures than their history—they also had some great ideas about running a business.

When we understand each other better, we can work together more effectively and have a better chance at success.

For example, if you're trying to sell something to someone who needs to speak your language or understand your culture, it will be difficult for them to buy from you. On the other hand, if they speak your language and understand your culture, there's a much higher likelihood that they will see why your product would be good for them and want it.

It can help us adapt our products and services to appeal more to people in other countries. For example, if we're selling products in Europe but need to know how Europeans shop or how they use their money, then we won't know how best to price our products or how much profit margin should be included with each sale. We will profit less from our sales, which means less money coming in overall!

Three Key Things we can Learn from the Ancient Greeks on Cultural Exchange

We have a lot to learn from the ancient Greeks regarding cultural exchange. In fact, there are three key things we can take away from their approach to this topic.

Greeks Learned to Adapt to Native Cultures

The ancient Greeks were known for their cultural exchange with the people of different countries they visited. They learned from each other's cultures, which helped them develop a better understanding of the world around them. When they went to foreign lands, they built strong relationships with the local people to form business deals. It helped them become more successful in their endeavours.

Greeks Built Strong Relationships With Foreign People To Form Business Deals

The people of Greece were the first Europeans to develop a strong relationship with the people of Africa, Asia, and Italy. They were very interested in making business deals with other countries and cultures. They often used these relationships as a way to gain new knowledge about different cultures, as well as make more money for themselves through trade agreements. Through these relationships, they developed a shared language and customs.

Greeks Broadened Their Horizons By Sailing Across The Mediterranean

The ancient Greeks were known for their ability to travel vast distances to explore new lands and cultures. It allowed them to learn about different ways of life, which resulted in a broader understanding of how people lived outside Greece at that time, resulting in greater cultural exchange between all parties involved (including Greece). They learned about foreign cultures by interacting with people from different parts of Europe and North Africa. This knowledge exchange allowed them to grow as individuals and as a society.

Adapting To Local Markets And Broadening Your Horizons Will Serve You Well

People often think of business as a zero-sum game where you can only gain by taking from someone else. But that's not the case.

The Greeks knew this well: they were the first to develop the concept of cultural exchange in business, and their approach served them incredibly well. They adapted their culture to local markets by expanding their empire and trade routes throughout the Mediterranean.

In doing so, they gained access to a larger market than if they had insisted on keeping their customs and traditions. They also earned the trust of local populations, who welcomed them with open arms—a faith that allowed them to expand their influence throughout the region even further.

When looking for ways to grow your business or make friends in a new city, remember the Greeks' example: try adapting your ideas and customs to broaden your horizons and create opportunities for others!

Benefits of Cultural Exchange in Today's Business

Cultural exchange is a powerful tool for business today. It can help with the following:

- You'll have a better understanding of the needs and wants of your customers, which means that you can create better products for them in the future.
- You'll better understand your customers, which will help you create more effective products and services.
- Your employees will have more open minds and diverse backgrounds, making them more likely to innovate.
- Cultural exchange helps build trust between countries and companies, strengthening bonds!
- You'll also get access to new markets and opportunities that would otherwise be inaccessible to a company that doesn't have an international presence. It can lead to higher revenues and profits for your company!

Although this era has ended, we still use their story to inspire today's business culture.

The Ancient Greek business culture was one of the most vital influences on today's business world. Although much of their mythology is steeped in magic and mysticism, their historical legacy has long passed. However, their story continues to inspire entrepreneurs today because, despite their differing belief systems, they were able to foster a shared vision and work tirelessly together for a better future. Not to say that there was no business before them, but rather, we can learn a great deal from their history in Greek times and how it has made its way into today's society.

Chapter 9

INNOVATION IN SHIPBUILDING

The ancient Greeks, surrounded by the deep blue expanse of the Mediterranean, had been true pioneers of shipbuilding. Their ships were more than mere gear for transport; they have been marvels of innovation, influencing each exchange and conflict. At the heart of their shipbuilding know-how had been the mythical triremes – smooth, speedy, and ambitious warships. Those vessels dominated the waves within the ancient globe with three rows of oars, leaving their mark on naval layout for generations. These grasp shipbuilders utilised an appropriate combination of substances, employing strong woods like cedar and oak, fortified with iron fittings. Their craftsmanship produced resilient and seaworthy vessels that enabled exploration, alternate, and navy conquest.

In this chapter, we will adventure into the sector of historic Greek shipbuilding by exploring the advancements in Greek shipbuilding and technology, innovations in Mediaeval European shipbuilding and modern shipbuilding innovations. We'll also examine the lessons modern businesses can learn from historic shipbuilding.

Advancements in Greek Shipbuilding and Technology

Historical Significance of Greek Shipbuilding Ancient Greece's maritime and naval prowess was long-lasting in ship layout and naval architecture. With their deep affinity for the sea, the Greeks have been no longer just craftsmen but also pioneers in shipbuilding. Their progressive techniques and engineering ingenuity have resonated through maritime history for centuries. Greeks diagnosed the significance of standardised creation methods, which emerged prominently in main port cities such as Athens, Rhodes, and Corinth. These strategies laid the foundation for constant green manufacturing, which resonates with current business practices.

The Trireme Warship

In ancient Greece, they built something remarkable – the trireme warship. This happened around 700 BC, and these ships were the superstars of their time. Do you know what made the trireme stand out? It was the way it moved. Instead of one row of oars, it had three, with around 170 rowers rowing together to make it go fast, up to 10 knots.

This speed and agility gave the triremes a huge advantage in sea battles. They could swiftly crash into enemy ships and sink them precisely, creating a strong naval force. Athens, one of the big Greek cities, had a fleet of these ships. They used them in famous battles like Salamis and Actium, where these warships changed the course of history. It would not be wrong to say that the trireme was the ultimate powerhouse of its time, revolutionising naval warfare.

Innovations in Design and Construction

The trireme's success stemmed from its oar system and ingenious design and construction innovations. Greek shipbuilders used progressive materials and techniques to maximise performance. A sturdy longitudinal keel provided structural integrity. Transverse ribs added support against stresses. The lightweight fir, linden, and pine hull improved speed and agility. Expert joinery with precise fittings created a seamless integration of components. Caulking materials and an outrigger beam enhanced seaworthiness for open-sea navigation. Standardised parts enabled rapid maintenance and repair. This combination of innovations showcased Greek shipbuilding craftsmanship and mirrored modern standardisation concepts. By leveraging materials science, structural design, and production techniques, Greek builders perfected the trireme as an unmatched naval vessel. The design and construction excellence behind this ancient technology remains inspirational today.

Spreading Greek Influence

The naval dominance of Greece in antiquity had profound results. It not only facilitated multiplied exchange around the Mediterranean and Black Seas but also had a profound influence on the colonies established by the Greeks. Greek ships, consisting of the renowned triremes, became vessels of tradition, information, and exchange. This effect extends beyond the confines of Greece, spreading to different historic civilisations.

Trade vessels, including the Syracusia grain provider, embarked on global voyages, carrying the simplest items and the maritime know-how of Greek shipbuilders. The effect of Greek advancements in shipbuilding extended to regions in which know-how became absorbed and delicate with the aid of different seafaring international locations, including the Romans, Egyptians, and Phoenicians.

Legacy of Greek Maritime Innovations

The legacy of Greek maritime improvements endures as a testament to the long-lasting effect of historic Greece on the maritime world. The Greeks valued maritime energy and diagnosed its ability to grow alternately, disseminate culture, and carry out naval missions. Over centuries, substances and production strategies have been delicate, laying the principles for broader European maritime traditions.

The Greek shipbuilding ingenuity of the past has paved the way for a generation of elevated exploration and alternate via sea. The enduring concepts of performance, adaptability, and craftsmanship stay as guiding lights for the modern maritime enterprise, a reminder that the instructions of history are by no means really out of date.

Greek Vessel Examples

Triremes: With its speedy oar-powered layout constructed for ramming enemy ships, the trireme stood as the dominant warship of historical Greece. Its legacy as an image of naval strength continues to encourage the contemporary maritime world.

Biremes: The bireme, an earlier warship prepared with two banks of oars, represented an evolutionary step towards the trireme. This progression displays the Greek dedication to refining their vessels.

Penteconters: Swift 50-oar galleys, referred to as penteconters, were widely used for scouting and patrols. Their speed and manoeuvrability made them critical to maritime surveillance and protection.

Syracusia: The Syracusia, a large grain service with novel lead sheathing, exemplified the achievement of Greek buying and selling vessels. Its voyages were a testament to the Greek maritime culture's impact on worldwide commerce.

Kyrenia Ship: The Kyrenia ship, a sophisticated service provider vessel, offers a precious time capsule demonstrating advanced joinery and creation techniques. Its renovation highlights the Greek dedication to craftsmanship.

Innovations in Mediaeval European Shipbuilding

The Cog Ship: A Cornerstone of Medieval Northern European Trade
Medieval Europe witnessed a large bounce in shipbuilding innovation with the emergence of cog delivery. These sturdy vessels became the workhorses of maritime trade, mainly in northern Europe. Cog ships have been designed to be robust transport vessels, perfectly suited for the growing needs of increasing maritime trade during the medieval period. These vessels had been instrumental in facilitating the movement of an extensive range of bulk cargoes throughout the buying and selling network of the Hanseatic League.

The cog's versatility allowed it to house an array of products, including bulk grain, salt, ore, and wine cargoes. Larger cogs, measuring over 25 meters long, had been well-suited for handling those hefty cargoes. This flexibility in shipment capacity made cogs invaluable for the change that flourished at some stage in this period.

Innovative Features of the Cog

The cog incorporated several innovative features that greatly improved its performance and utility as a medieval sailing vessel. One key innovation was the sternpost rudder, which provided greater directional control than the steering oars used on earlier ships. This made navigation safer and more precise.

Another innovation was using multiple masts with lateen sails on each mast. This sail configuration allowed cogs to sail closer towards the direction of the wind, increasing their manoeuvrability and sailing efficiency.

Finally, the hull design, featuring a rounded shape and a very wide beam, maximised the internal cargo capacity of the cog. This was crucial for the bulk transport of goods and materials. The expanded cargo space and improved sailing performance made cogs excellent cargo carriers.

Together, these innovations made the cog a more adaptable, high-performing sailing vessel ideal for navigation and large-scale shipping operations during medieval times.

Transition to Standardised Skeleton-First Construction

During the medieval period, shipbuilding techniques had evolved, transitioning from shell-first creation to more standardised skeleton-first techniques. This shift was specifically mentioned in the manufacturing of cogs. Using pre-cut rib frames and different standardised components allowed for more green batch assembly line manufacturing. As a result, the construction of cogs became prolific, using early marine industrialisation.

This transition to standardised skeleton-first production has expedited the shipbuilding procedure and contributed to the general improvement of cog layout and overall performance. It set the level for the mass manufacturing of those modern vessels, which were instrumental in shaping the maritime change panorama of medieval Europe.

Venice and the Sea Trade Monopoly

Venice, a maritime powerhouse of the medieval generation, leveraged the improvements of cogs to extend its sea trade monopoly throughout Europe and Asia. Venetian traders and shipbuilders diagnosed the capacity of these vessels for the green transportation of their coveted goods, along with Venetian glass, textiles, and spices.

Bulk shipment cogs were pivotal in exporting those precious commodities to remote markets. Armed service provider cogs, mainly designed for protection, observed and safeguarded these lucrative convoys throughout their trips. The efficiency of Venetian shipyards, exemplified by the renowned Arsenal shipyard in Venice, further proved the effectiveness of cog production and contributed to Venice's dominance in the maritime trade region.

Age of Sail: Fast Clipper Ships

Extreme clipper ships represented the pinnacle of sailing technology. With long, narrow, strongly curved hulls and vast sail plans, they were designed purely to maximise speed while transporting cargo and passengers. Clippers raced to set speed records between Europe, America, and Asia.

Streamlined hulls were crafted using lightweight tropical hardwoods reinforced internally with iron knees and frames. Smooth copper, zinc, or lead sheathing reduces drag and marine fouling. Hulls tapered sharply for peak hydrodynamic performance. Innovative sail technologies like 3–4 masted rigs enabled carrying thousands of square feet of canvas. Gaffs, multiple headsails and topsails per mast, and rigging refinements provided immense sailing power and control. Navigation instruments were also improved. Clippers generated immense fortunes by shrinking long-haul maritime routes. They rapidly expanded global trade, especially for high-value goods like tea, opium, and wool. Clipper crews also transported millions seeking new frontiers and opportunities.

Steam power and steel hulls

The Industrial Revolution enabled the transition from sail to coal-fired boiler steam propulsion in the 1800s. Early low-pressure paddle steamers plied rivers and coasts before compact steam engines made reliable ocean travel practical. Iron and steel became essential shipbuilding materials, enabling increased hull dimensions, capacity, and structural integrity as garboard plates replaced frames. Year-on-year, ships grew larger and more complex.

Reciprocating steam engines, screw propellers, and condenser innovations substantially boosted available propulsive power while increasing efficiency and reliability. Screw propellers offered superior manoeuvrability over paddle wheels, along with greater economy. These innovations enabled regular, safe transoceanic transportation. Steel-hulled steamships carried millions of passengers and tonnes of vital cargo across the globe, facilitating trade and migration. Lusitania and Mauretania highlighted progress.

Modern shipbuilding innovations

Modern ships leverage advanced propulsion systems, hydrodynamic hull designs, digital navigation technologies, and simulation software to optimise performance. Efficient engines, lightweight materials, and alternative fuels improve propulsion. Streamlined hulls, bow designs, and optimised components reduce drag and improve seakeeping. Integrated digital systems enhance navigational safety and precision. Simulation modelling and analysis enable builders to optimise designs for efficiency, capacity, sustainability, and safety.

Lessons for Modern Businesses from Innovation in Marine Transportation (Greek Shipbuilding)

Following are the lessons that modern businesses can learn from Greek shipbuilding innovations:

Embracing Innovation for Industry Advancement

The history of maritime transportation, exemplified by the modern spirit of Greek shipbuilding, provides precious training for cutting-edge corporations. One of the foremost classes is the mammoth value of decisively embracing innovation and applying new technologies to propel the enterprise ahead, specifically at some point of disruptive alternate. The historic Greeks understood that embracing new ideas and technologies was key to achieving maritime dominance. Similarly, current agencies must understand that innovation isn't a choice but a necessity for staying competitive and relevant in the ever-evolving panorama of the 21st century.

Cultivating an innovation mindset

Innovation is not restrained to technological breakthroughs; it extends to cultivating a typical innovation mindset within agencies. Maritime history demonstrates that cultivating ingenuity and an innovation mindset allows groups to substantially improve key aspects of their operations, which include speed, efficiency, protection, sustainability, and universal price creation. Just as Greek shipbuilders constantly sought to beautify their vessels and operational strategies, present-day corporations need to adopt a culture of innovation that encourages personnel to think creatively, discover possibilities for improvement, and force changes that pressure development.

Adaptation to Changing Conditions

One of the most essential pieces of maritime history is the desire for constant edition. Successful shipbuilders in the ancient world and within the cutting-edge generation must adapt their vessels and operations to fulfil evolving financial conditions, environmental requirements, competitive dynamics, and client necessities. The ability to pivot, reply to marketplace needs, and proactively alter adjustments within the enterprise is paramount. In an international environment in which globalisation and technological improvements are swiftly reworking commercial enterprise landscapes, the adaptability of organisations determines their resilience and long-term success.

Fostering a Culture of Innovation and Excellence

To maintain a robust maritime legacy and ensure enduring competitive advantages, present-day agencies must proactively foster a culture of innovation, engineering excellence, and layout management. Because the Greek shipbuilders left a lasting legacy with their revolutionary designs, cutting-edge organisations can build their legacies by prioritising innovation and excellence in their respective fields. By encouraging a commitment to innovative questioning, placing excessive requirements for engineering and layout, and always pushing the boundaries of what is feasible, companies can secure their position as enterprise leaders and contribute to a legacy of development.

Leveraging Tradition to Drive Progress

The maritime records of Greek shipbuilding teach modern corporations the treasured lesson of leveraging the way of life to drive development. While embracing innovation is important, it need not come at the price of dismissing mounted knowledge and practices. Greek shipbuilders built upon centuries of amassed know-how, refining their methods while respecting centre standards. Similarly, modern organisations can benefit from respecting and integrating established pleasant practices while pushing innovation bounds.

By applying these lessons, agencies can navigate the demands of the contemporary enterprise world with self-assurance and motive, just as the Greeks did in their pursuit of maritime excellence.

Summarising this chapter, throughout history, major inflexion points in shipbuilding innovation have revolutionised maritime capabilities and global trade. From triremes improving naval power to clippers optimising sailing speed and efficiency to the complexity of modern cruise ships, each advancement enabled new possibilities. Fundamentally, the marine industry relies on continuous innovation across technology, design, materials, processes, and business models. Meaningful progress depends on pushing boundaries. By leveraging incremental improvements and disruptive innovations, shipbuilders can shape the industry's future evolution. New technologies can further optimise performance, efficiency, sustainability, and safety. Although competitive forces may shift, human ingenuity is an enduring renewable resource for progress. With integrated innovation strategies, the marine industry's brightest years likely still lie ahead. The future remains ours to build through creative problem-solving and persistence in overcoming challenges. The next revolutionary innovation could be right around the corner.

CONCLUSION

From ancient Greece's contributions in pioneering the coinage system to advancing ship design, particularly the trireme, have transcended time, traders and shipbuilders of ancient Greece undeniably etched their footsteps in the history of ancient civilisations. It's the enduring value of their teachings that keeps their memory alive even after centuries.

We discover remarkable relevant ideas today when we examine how the ancient Greeks conducted trade by sea and with one another. These ideas encompass building durable ships, exploring new trade routes and establishing colonies. Moreover, they involve knowing where to market your goods, employing the right currency, engaging with middlemen, practising fairness, engaging with diverse cultures, and enhancing transportation efficiency.

By applying these age-old principles in modern business, we can enhance our shipping and trading practices, foster business growth, and seize opportunities in the global market. It's akin to gleaning wisdom from the past to achieve better outcomes in the present. The legacy of ancient Greece persists as a wellspring of knowledge for contemporary endeavours.

FINAL SUMMARY

While Greek mythology's history and captivating tales are not as old as time, they are still quite old but relevant in our modern-day society. Some famous figures known to many, such as Zeus, Hades and Poseidon, are just a small part of the Greek empire.

The Greeks were dominant traders who understood the significance of having a well-conditioned navy for political and trading purposes. Their use of amphoras is another fascinating aspect of Greek trade, as these containers were used to carry and categorise products like grain, wine and oil.

With an extensive trading network, their naval force had to remain formidable to defend trade routes with warships that posed as security detail for cargo ships to reduce the chances of piracy on the high seas.

In an effort to resolve internal tensions and reduce the risk of civil war, the Greeks sought to colonise various countries along trade routes. In fact, by applying these age-old principles in modern business, we can enhance our shipping and trading practices, foster business growth, and seize opportunities in the global market.

www.ingramcontent.com/pod-product-compliance
Lightning Source LLC
Chambersburg PA
CBHW080900030426
42335CB00018B/2414